I0505401

ILLINOIS DMV PERMIT TEST GUIDE

Drivers Permit & License Study Book

With Success Oriented Questions & Answers for Illinois DMV written Exams 2020

LISA PARKS

All Rights Reserved. No Contents of this book may be reproduced in any way or by any means without duly certified consent of the publisher, with exception of brief excerpt in critical reviews and articles.

Introduction

It isn't news that the DMV average knowledge test pass rate in the United States is a terrible 49%. However, some persons tend to depend on their states driver's manual only for their Exams and get to the DMV overconfident but unprepared. Don't allow this to happen to you, it can be different. It's my belief that people need to know what to expect on their DVM DL exam so as to prepare very well. Here comes a well prepared question and answers study manual/book that will increase your chances of passing and gives you the peace of mind so you will clear the official exam on your first attempt.

This manual will give you every necessary help you will ever need to pass the DMV Exam irrespective of the part of the States you presently live in! However, without any exaggerations, I'm supper sure that if you can give-in a little time to studying this manual it would certainly serve as a standard guide towards ensuring that you pass your DMV with ease. Taking these practice permit tests will help you to get acquainted with the real test thereby, making your anticipated success in the exams a reality.

With this manual, there is really no need to be afraid as the questions contained in it are close enough to what you will see and be tested on in the real test. This test manual has different sections of what you will be tested on based on experience. And there are many questions and answers in it, which will give you an in-depth knowledge as well as sufficient preparation for the real test. The questions cut across; defensive driving, road signs/markings and turnings. It also includes some questions on braking, steering techniques, and skid controls, and much more. As a matter of necessity, you are strongly advised to do well to repeat each test contained herein until you can achieve a consistent score of about 90% and above. In this book you will learn the exact things that those people who pass on their first attempt always do:

General driving practice test

Defensive driving test

Teen drivers test

Road signs and traffic control seen on the highway, streets and walkway,

Please note that the correct answers to the test questions are contained at the back of the book,

Test on General Driving Experience

1.   Driving and doing any of the following; eating, drinking, angry, ill, and texting are all examples of?

a)   Safe Driving.

b)   Defensive Driving.

c)   Talented Driving.

d)   Distracted Driving.

2.   Driving closely behind a vehicle is known as what?

a)   Drafting.

b)   Towing.

c)   Tailgating.

d)   None of the above.

3.   Keep your eyes on the road with a firm grip on the stirring, quickly engage neutral pull off the road when it's

safe to do so, and turn the engine off. These are things you do when there is?

a)   A tire blowout.

b)   Power failure.

c)   Headlight failure.

d)   A stuck gas pedal.

4.   The cause of most rear end collisions is?

a)   Failing to inspect the vehicle.

b)   Not checking your mirrors.

c)   Talking to passengers.

d)   Following too closely.

5.   When at a railroad crossing that does not have signals what is the most appropriate action to take?

a)   Come to a complete stop.

b)   Slow down and be prepared to stop.

c)   Speed to get across the tracks quickly.

d) None of the above.

6. You may use your horn to alert drivers that they have made an error.

a) True

b) False

7. What would you do if a railroad crossing has no warning devices?

a) Stop at once within 15 feet of the railroad crossing.

b) Increase your speed and cross the tracks very quickly.

c) Slow down and then proceed with caution.

d) All crossings have a control device.

e) None of the above.

8. When is it really legal to drive a vehicle in a bicycle lane?

a) When preparing to turn. For less than 200 feet

b) If your hazard lights are on.

c) If there are no bikes.

d) It is never legal driving a vehicle in a bicycle lane.

9. How fast may one drive when driving on a highway posted for 65 mph and the traffic is traveling at 70 mph.

a) As quickly as the speed of traffic.

b) Between 65 mph and 70 mph.

c) As fast as the momentum needed to pass other traffic.

d) Not faster than 65 mph.

10. Keeping your eyes on the road, quickly shifting to neutral, pulling off the road when it is safe to do so, and turning off the engine are the procedures for which of the following ?

a) A stuck gas pedal.

b) A power failure.

c) Brake failure.

d) None of the above.

11. It isn't necessary at all to signal before you exit a freeway or pull away from a curb?

a) True

b) False

12.   Once you miss your exit on a freeway, it is legal to stop and back up on the shoulder.

a) True

b) False

13.   In what situation is it OK to back up on the highway?

a)   If you miss your exit.

b)   To go back to see an accident.

c)   To pick someone on the side of the highway.

d)   It is really never ok to back up on the highway.

14.   The best way to enter a freeway very smoothly is to accelerate on the entrance ramp so as to match the speed of freeway traffic in the right lane.

a) True

b) False

15.   Which of these statements about large trucks is true?

a) Trucks usually travel slower than cars.

b) Trucks often make wide right turns.

c) Trucks take longer to stop than cars.

d) All of the above.

e) None of the above.

16. What do blind spots mean?

a) Blind spots are places for blind people to cross an intersection.

b) Blind spots are dots only seen by drivers who have been drinking.

c) Blind spot is an area near the left and right rear corners of the vehicle which cannot be seen in the rear-view mirrors.

d) Blind spot are spots seen after staring into oncoming headlight at night.

e) None of the above.

17. What is actually the minimum safe following distance under most conditions,?

a) The recommended following distance under most conditions is a minimum of 2 seconds.

b) The minimum of 4 seconds is the recommended following distance in most conditions.

c) A minimum of 5 seconds is the recommended following distance in most conditions.

d) 6 seconds minimum is the recommended following distance under most conditions.

e) None of the above.

18. When is it permissible to drive faster than the posted speed limit?

a) When being followed too closely.

b) To be keeping pace with the flow of traffic.

c) If you have an emergency.

d) It is never permissible.

19. One can cross a solid yellow line to do which of the following?

a) To do a U turn on the highway.

b) To turn into a driveway only if it is safe to do so.

c) To pass on the highway.

d) To get a good view of the road ahead.

20. When moving at 50 mph it will take approximately how many feet to react to an object on the road and stop your vehicle?

a) 200 feet.

b) 300 feet.

c) 400 feet.

d) 500 feet.

21. You should dim your high beams whenever you come within ____ feet of an oncoming vehicle?

a) 200

b) 300

c) 400

d) 500

22. A solid white line signifies what part of the traffic lane on a road?

a) It separates the lanes of traffic moving in opposite directions. Single white lines can also mark the right edge of the pavement.

b) It separates lanes of traffic moving in same direction. Single white lines also mark the left edge of the pavement.

c) It separates lanes of traffic moving the same direction. Single white lines can also mark the right edge of the pavement.

d It separates lanes of traffic moving the opposite way. Single white lines can also mark the left edge of the pavement.

e) None of the above.

23. While driving a motor vehicle, both hands should be on the steering wheel at all times except you are texting.

a) True

b) False

24. If ever a tire blows out, you should?

a)   Hold the steering wheel very firmly and gradually bring the vehicle to a halt without using the gas pedal.

b)   Apply the brakes firmly.

c)   Shift to neutral and apply the brakes.

d)   Speed up to gain stability, and pull over.

e)   None of the above.

25.   A motorist when approaching a bicyclist should?

a)   Hurry and pass him.

b)   Proceed as normal.

c)   Swerve into the opposite lane.

d)   Exercise extreme caution.

e)   Stay behind following at a safe distance

26.   Talking on cell phone may increase your chances of being in a crash by as much as four times.

a)   True

b)   False

27. Driving relatively much slower than the speed limit in normal conditions can do what?

a) It can decrease the chance of an accident.

b) It does not change anything.

c) It can increase driver safety.

d) It can increase the chances of an accident.

28. Trying to move too fast from a stop could cause which of the following?

a) Battery strain.

b) Stuck gas pedal.

c) Spinning drive wheels.

d) Excessive smiling.

e) None of the above.

29. What type of vehicles must stop at all railroad crossings?

a) All vehicles.

b) Recreational vehicles.

c) School buses and passenger buses.

d) 18 wheelers.

e) All of the above.

30. Passing on the right is permissible when it is safe and the driver of the other vehicle is making a left turn

a) True

b) False

31. Motorcycles cannot stop as quickly as any other vehicle can?

a) True

b) False

32. Whenever you pass a vehicle traveling in the same direction, you are expected to pass on the left?

a) True

b) False

33. To pass on the right is permissible though not ideal on a one-way road, streets and highways that are marked for two or more lanes of traffic moving in the same direction.

a) True

b) False

34. You have to give-in to traffic on your right already in a roundabout.

a) True

b) False

35. Hitting any vehicle moving in the opposite direction is better than hitting one moving in the same direction

A) True

b) False

36. There are some circumstances where it is legal to double park

a) True

b) False

37. What would you do at an intersection with a flashing red light?

a) Come to A full stop, and go whenever it flashes green and is safe to do so.

b) Come to a full stop, and go as soon as you can do so.

c) Slow down and yield to vehicles already in the intersection.

d) Come to a full stop, then go whenever it turns solid green

38. When should safety belts be worn?

a) At all times while driving and as a passenger.

b) Only when driving on curvy roads.

c) Only when riding in the back seat.

d) Only when driving on the freeway.

e) None of the above.

39. When approaching any intersection with traffic control signals that are not working, you should treat it as you would a 4-way yield sign

a) True

b) False

40. What does a flashing yellow light indicate?

a) Stop, then proceed with caution.

b) Proceed with caution.

c) Pedestrian crossing.

d) School Crossing.

e) None of the above.

41. Whenever an emergency vehicle is approaching you with a siren and flashing lights it is ok to continue at the same speed if there is another lane open.

A) True

b) False

42. Whenever you see a stopped vehicle on the side of the road what would you do?

a) Stop and offer assistance.

b) Slow down and proceed with caution.

c)   Sound the horn to let them know that you are about to pass them.

d)   Notify emergency services.

e)   None of the above.

43.   Quickly tapping the brake pedal 3 or 4 times would help to let those driving behind you know that you are about slowing down.

a)   True

b)   False

44.   When passing, you may exceed the posted speed limit only when passing a group of cars.

a)   True

b)   False

45.   When could you signal if you plan to pull into a driveway just after an intersection?

a)   After you cross the intersection.

b)   Before you cross the intersection.

c) When you start your turn.

d) In the middle of the intersection.

46. When may you legally block an intersection?

a) During rush hour traffic.

b) If you entered an intersection when the light was green light.

c) You cannot legally block an intersection.

d) When the light is yellow.

e) None of the above.

47. Where are ramp meters usually located?

a) In parking garages.

b) On Loading docks.

c) On highway exit ramps.

d) On highway entrance ramps.

e) None of the above.

48. When is it really ok to drive faster than the posted speed limit?

a)   When being followed too closely.

b)   To keep pace a with the flow of traffic.

c)   If you have an emergency.

d)   It is never ok.

49. Name some likely places where you may find slippery spots on the road.

a)   In corners and at stop signs.

b)   In shady spots as well as on overpasses and bridges.

c)   In tunnels and on hills.

d)   Near large bodies of water.

e)   None of the above.

50. After drinking alcohol, a cold shower or coffee will lower your blood alcohol content.

a)   True

b)   False

51.   There are some situations when it is legal to double park .park.

a)   True

b)   False

52.   Safety belts would help you keep control of your car.

a)   True

b)   False

53.   What would you do if you are at an intersection and you hear a siren?

a)   Stop and do not move until after the emergency vehicle has passed.

b)   Continue through the intersection, and pull over to the right side of the road then when it's safe and stop.

c)   Back pull over to the right side of the road then stop.

d)   None of the above.

54.   The state examiner would check the person's vehicle before beginning the driving test to:

a) Ensure the vehicle has all the necessary equipment.

b) Make sure that the vehicle is in a safe operating condition.

c) Check for cleanliness.

d) A and C.

e) A and B.

55. One of the basic things to remember about driving at night or in fog is to?

a) Be ready to brake more quickly.

b) Watch for cars at intersections.

c) Drive within the range of your headlights.

d) Use your high beams at all times.

e) None of the above.

56. It is very ok to continue at the same speed when an emergency vehicle approaches you with flashing lights and a siren if there is any other lane open.

a) True

b) False

57. Which of the following about littering while driving is true?

a) It may cause a traffic accident.

b) Is against the law.

c) It could lead to large fines up to and including jail time.

d) All of the above.

e) None of the above.

58. Which of these influences the effects of alcohol in the body?

a) How much time between each drink.

b) The body weight of a person.

c) The amount of food in the stomach.

d) All of the above.

e) None of the above.

59. When at a railroad crossing that does not have signals what in your opinion is the correct action to take?

a) Come to a complete stop.

b) Slow down and be prepared to stop.

c) Speed to get across the tracks quickly.

d) None of the above.

60. To keep a steady speed and signaling in advance when slowing down or turning will help maintain what

a) A safe distance ahead of your vehicle.

b) A safe distance behind your vehicle.

c) A safe distance next to your vehicle.

d) All of the above.

e) None of the above.

61. Which of the following would be most effective in avoiding a collision?

a) Keeping your lights on at all times.

b) Wearing a seat belt.

c)   Driving in daytime hours only.

d)   Keeping cushion of space at all times.

e)   Driving slow at all times

62.   In driving a roundabout, the same general rules apply as for maneuvering through any other type of intersection.

a)   True

b)   False

63.   When a signal light turns green, what would you do?

a)   Accelerate as quickly as possible.

b)   Yield to pedestrians.

c)   Count two seconds before accelerating.

d)   Do not move until another driver signal you to go.

e)   None of the above.

64.   While operating a motor vehicle, your both hands should be on the steering wheel at all times except you are texting

a)   True

b)   False

65.   How fast would you legally drive if driving on a highway posted for 65 mph and the traffic is traveling at 70 mp?

a)   As fast as the speed of traffic.

b)   Between 65 mph and 70 mph.

c)   As fast as the necessary speed needed to pass other traffic.

d)   No faster than 65 mph.

66.   When you park on an uphill grade, which way should you turn your wheels?

a)   Left

b)   Right

c)   Straight

d)   None of the above

Test on Knowledge of Road Signs &Traffic Control

1. A cross-buck or white X shaped sign that has Railroad Crossing written on it has the same meaning as a stop sign.

a) True

b) False

2. The moment you have a yellow line on your right and white line on your left are you going the wrong way?

a) Yes

b) No

3. Except prohibited by a sign, when may one turn left at a red light?

a) In an emergency.

b) From two way road to a one way road.

c) From a one way road to another one-way road.

d) Never.

4. What does a red painted curb indicate?

a) Loading zone.

b) Reserved for passenger drop off or pick up.

c) No parking or stopping.

5. Pavement line colors show if you are on a one-way or two-way roadway.

a) True

b) False

6. When approaching a road construction work zone, what should you do?

a) Close the distance between your vehicle and any vehicle ahead of you.

b) Prepare to slow down or stop.

c) Prepare to watch the ongoing work.

d) Put on hazard lights to warn other drivers.

7. What do white painted curbs signify?

a) Loading zone for freight or passengers.

b) No loading zone.

c) Loading zone for passengers or mail.

d) No loitering.

e) None of the above.

8. What do an orange sign imply?

a) State highway ahead.

b) Merging lanes ahead.

c) Construction work ahead.

d) Divided highway road ahead.

9. When can one make a left turn at a green light?

a) Only if there is a green arrow.

b) Only on a city streets.

c) Only after yielding to oncoming traffic.

d) Only on one-way streets.

10. When you are on the highway, how far ahead should you look?

a) 1 city block.

b) 1 quarter mile.

c) 1 half mile.

d) 1 mile.

11. What color lines divide lanes of traffic going in opposite directions?

A) White

b) Yellow

c) Orange

d) Red

12. If a traffic light turns from green to yellow as you approach an intersection, what should you do?

a) Stop, even if in the intersection.

b) Speed up so as to beat the light before it turns red.

c) Keep going at your current speed.

d) Stop before the intersection.

13. When you merge with traffic, at what speed should you try to enter traffic?

a) A slower speed than traffic.

b) The same speed as traffic.

c)   A faster speed than traffic.

d)   As fast as you can go.

14.   Multiple lanes of travel going in the same direction are usually separated by lane markings of what color?

a)   Red

b)   Yellow

c)   White

d)   Orange

15.   A cross-buck or white X shaped sign that says Railroad Crossing on it has the same meaning as a yield sign

a)   True

b)   False

16.   What do a yellow sign mean?

a)   State highway ahead.

b)   A special situation or a hazard ahead.

c)   Construction work ahead.

d)   Interstate sign.

17.   Any pedestrian using a white or a red-tipped white cane is usually what?

a)   A policeman.

b)   A construction worker.

c)   A blind person.

d)   A crossing guard.

e)   None of the above.

18.   What is really the correct left-turn hand signal?

a)   Hand and arm extended downward.

b)   Hand and arm extended out.

c)   Hand and arm extended upward.

d)   Hand and arm is extended upward with middle finger extended upward

19.   You are not required to make a full stop in which situation?

a)   At a steady red traffic signal.

b)   At a flashing yellow traffic signal.

c)   At a stop sign.

d)   At a flashing red traffic signal.

20.   What do broken white line on the highway signify?

a)   A broken white lines separates two lanes traveling in opposite directions. You may cross this line when changing lanes, once you have signaled and it is safe to do so,

b)   A broken white line separates two lanes that travels the same direction. Do not cross this line.

c)   A broken white line separates is two lanes traveling the opposite direction. Do not cross this line.

d)   A broken white line is often used to separate two lanes traveling the same direction. You may cross this line when changing lanes once you have signaled and it is safe to do so,

e)   None of the above.

21.   You could cross a double yellow line to pass another vehicle if the yellow line next to you is what

a)   A solid line.

b)   A thinner line.

c)   A broken line.

d)   A thicker line.

e)   None of the above.

22.   If you have a green light, and traffic is backed up into the intersection, what should you do?

a)   Enter the intersection and hope that traffic clears before the light changes.

b)   Wait for traffic to clear before you enter the intersection.

c)   Try to go around the traffic.

d)   Sound your horn to clear the intersection

23.   When you have a green light, and traffic is backed up into the intersection, what do you do in that instance?

a)   Enter the intersection and hope the traffic clears before the light changes.

b)   Wait until traffic clears before entering an intersection.

c)   Try to go around the traffic.

d)   Sound your horn to clear the intersection.

e)   None of the above.

24.   Put the following signs in their proper order from left to right.

a)   Yield, school zone, construction.

b)   School zone, construction, no passing.

c)   Pass with care, construction, yield.

d)   No passing, slow moving vehicle, yield.

e)   No passing, construction, yield.

25.   Passing is permissible in either direction if there are two solid yellow lines in the center of the road when?

a)   When following a slow truck.

b)   Only when it is safe.

c)   On US highways only.

d)   Passing is never permitted.

26.   What do a yellow sign indicate?

a)   State highway ahead.

b)   A special situation or a hazard ahead.

c)   Construction work ahead.

d)   Interstate Sign.

e)   None of the above.

27.   A white square or a rectangular signs with white, red, or black letters or symbols are usually what kind of signs?

a)   Destination

b)   Service

c)   Route

d)   Reference

e)   Regulatory

f)   Warning

28. The Yellow diamond signs with black letters or symbols are what kind of signs?

a) Destination

b) Service

c) Route

d) Reference

e) Regulatory

f) Warning

29. Both of these signs shows that you are entering a School Zone.

a) True

b) False

30. Multiple lanes of travelling in the same direction are separated by lane markings of what color?

a) Red

b) Yellow

c) White

d) Orange

31. What do an orange sign mean?

a) State highway ahead.

b) Merging lanes ahead.

c) Construction work ahead.

d)   Divided highway ahead.

e)   None of the above.

32.   A solid white line shows what part of the traffic lane on a road?

a)   It separates lanes of traffic moving the opposite directions. Single white lines could also mark the right edge of the pavement.

b)   It separates lanes of traffic moving the same directions. Single white lines could also mark the left edge of the pavement.

c)   It separates lanes of traffic moving the same direction. Single white lines could also mark the right edge of the pavement.

d) It separates the lanes of traffic moving in the opposite direction. Single white lines may as well mark the left edge of the pavement.

e) None of the above.

33. If a traffic light change from green to yellow as you approach an intersection. What should you do?

a) Keep going at your current speed.

b) Stop before the intersection.

c) Stop, even if in the intersection.

d) Speed up to beat the traffic light before it turns red.

e) None of the above.

34. What do white painted curbs indicate?

a) Loading zone for freight or passengers.

b)   No loading zone.

c)   Loading zone for passengers or mail.

d)   No loitering.

e)   None of the above.

35.   At an intersection what should you do with a flashing

red light, ?

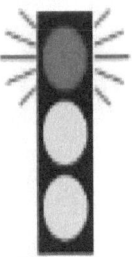

a)   Come to a stop, then go when safe to do so.

b)   Come to a stop, then go when it flashes green.

c)   Slow down and yield to the vehicles already in the intersection.

d)   Come to a stop, then go when it turns solid green.

36.   When more than one vehicle arrives at the same time at a four way stop, which vehicle should go first?

a)   The first one that attempts to go.

b)   The vehicle on the left.

c)   The vehicle on the right.

d)   None of the above.

37.   A white square or a rectangular signs with white, red, or black letters or symbols are usually what kind of signs?

a)   Destination

b) Service

c) Route

d) Reference

e) Regulatory

f) Warning

38. When is it appropriate to obey instructions from school crossing guards?

a) Only during school hours.

b) Only if you see children present.

c) Only if they are licensed crossing guards.

d) Only at a marked school crosswalk.

e) At all times.

39. What does this hand signal signify?

a) Left turn.

b) Right turn.

c) Stop or slowing down.

d) Backing.

e) None of the above.

40. What do you think this hand signal mean?

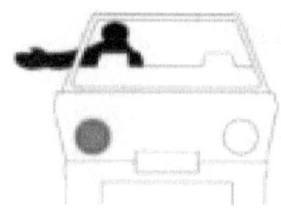

a) Left Turn.

b) Right Turn.

c) Stop or Slowing Down.

d) Backing.

e) None of the above.

41. Traffic Light Meaning: Stop, yield to the right-of-way, and go when it is safe.

a) Red arrow.

b) Steady yellow.

c) Flashing yellow.

d) Green arrow.

e) Flashing red.

f) None of the above.

42. A Pentagon shaped sign signify which of the following?

a) No Passing Zone.

b) Railroad Crossing.

c) School Zone

d) Yield.

e) Stop.

43. The picture below shows an improper hand placement on the steering wheel.

a) True

b) False

44. What does this hand signal indicate?

a) Left turn.

b) Right turn.

c) Stop or slowing down.

d) Backing.

e) None of the above.

45. Traffic Light Meaning: You must not go in the direction of the arrow until the light is off and a green arrow or a green light comes on.

a) Red arrow.

b) Steady yellow.

c) Flashing yellow.

d) Green arrow.

e) Steady green.

f) None of the above.

46. What does this orange sign mean?

a) One handed man with lunch.

b) Beware of car-jacking.

c) Construction flagger ahead.

d) Watch for pedestrians.

e) None of the above.

A   B   C

47.   Which sign above is a guide sign?

a)   A.

b)   B.

c)   C.

d)   None of the above.

48.   Which of these best describes this sign's meaning?

a)   Road spins.

b)   Drive in circles.

c)   Roundabout ahead.

d)   Circular highway.

e)   None of the above.

49.   What does this sign mean?

a)   Road splits

b)   Yield

c)   Merge

d)   Divided highway

e)   Right lane ends

A   B   C

50.   Which of the signs above is a warning sign

a)   A.

b) B.

c) C.

d) None of the above.

A   B   C

51. Which sign above means passing permitted

a) A.

b) B.

c) C.

d) None of the above.

52. Which of these best describes this signs meaning?

a) Construction worker.

b) Pedestrian crossing.

c) School crossing.

d) Jogging path.

e) None of the above.

53. This sign means which of the following?

a) Look both ways.

b) Entering 2 way traffic ahead.

c) You can go forward or reverse.

d) One way traffic.

e) None of the above.

54. What does this sign mean

a) While the light is red even if the road is clear.no turning

b) No turning on green or red.

c)   You must ensure the road is clear and if the light is red, then you can turn.

d)   None of the above.

55.   What do this sign mean?

a)   4 way stop ahead.

b)   An intersection of roads ahead.

c)   Divided roadway.

d)   Yield ahead.

e)   None of the above.

56.   What do this bicycle sign mean?

a) Bicycles race starts here.

b) No bicycles.

c) Bicycle crossing.

d) Bicycles must park by the arrow.

e) None of the above.

57.   Which of these options best describes this sign?

a) Road splits

b) Yield

c) Merge

d) Divided highway

e) Right lane ends

58.   What kind of sign is this?

NO
TURN
ON RED

a)   Warning sign.

b)   Regulatory sign.

c)   Construction sign.

d)   Guide sign.

e)   None of the above.

59.   This sign means which of the following?

a)   Look to your left.

b)   Stop if turning left.

c)   Curve to the left.

d)   Sharp turn to the left.

e) None of the above.

60. When should a driver obey instructions from school crossing guards?

a) Only during school hours.

b) Only if you see children present.

c) Only if they are licensed crossing guards.

d) Only at a marked school crosswalk.

e) At all times.

61. When do you obey a construction flogger's instructions?

a) Only if you see it necessary to do so.

b) If they do not conflict with existing signals.

c) If they are wearing a state badge.

d) At all times in construction zones.

e) None of the above.

62. When the law enforcement is directing you to drive through a red light, what should you do?

a)   Never ignore traffic signals.

b)   Drive through the red light.

c)   Wait for the light to turn green.

d)   Pull over and stop.

63.   What does a flashing yellow light indicate?

a)   Pedestrian crossing.

b)   School Crossing.

c)   Stop, then proceed with caution.

d)   Proceed with caution.

e)   None of the above.

64.   If you are planning to pull into a driveway immediately after an intersection. When should you signal?

a)   After you cross the intersection.

b)   Before you cross the intersection.

c)   When you start your turn.

d)   In the middle of the intersection.

65.   What would you do if a railroad crossing has no warning devices?

a)   Stop within 15 feet of a railroad crossing.

b)   Increase your speed, cross the tracks quickly.

c)   Slow down and then proceed with caution.

d)   All crossings have a control device.

e)   None of the above.

Test on Lane Control

1.   Which lane would you end up in after completing a turn when turning from one of three turn lanes?

a)   The right lane if clear.

b)   The lane you started in.

c)   The left most lane.

d)   Always the middle lane.

2.   Broken yellow lines separates lanes of traffic going in the same direction

a)   True

b)   False

3.   Whenever you pass a vehicle traveling in the same direction, you are expected to pass on the left?

a)   True

b)   False

4.   If you are traveling 47 mph on a highway with a speed limit of 55 mph which lane should you be in?

a)   The far left lane.

b)   The carpool lane.

c)   The far right lane.

d)   The middle lane.

e)   The bicycle lane.

5. Which vehicle goes first if more than one vehicle arrives at the same time at a four way stop?

a) The first one that attempts to go.

b) The vehicle on the left.

c) The vehicle on the right.

d) None of the above.

6. Passing is permissible in either direction when there are two solid yellow lines in the center of the road when?

a) When following a slow truck.

b) Only whenever you are sure it is safe.

c) On US highways only.

d) Passing is never permitted

7. You can cross a solid yellow line to do which of the following?

a) To make U turn on the highway.

b) To turn into a driveway when it is safe to do so.

c) To pass on the highway.

d)   To get a clearer view of the road ahead.

8.   If you want to turn to the left at an intersection but the oncoming traffic is heavy, what would you do?

a)   Wait at the crosswalk while traffic clears.

b)   Wait at the center of the intersection for the traffic to clear.

c)   Start your turn forcing others to stop.

d)   None of the above.

9.   Which of these statements is correct?

a)   A solid or dashed yellow line shows the left edge of traffic lanes going in your direction.

b)   A solid or dashed yellow line shows the right edge of traffic lanes going in your direction.

10.   A driver turning left at an intersection is expected to yield to what?

a)   Vehicles approaching from the opposite direction.

b)   Pedestrians, bicycles and vehicles, approaching from the opposite direction.

c)   Pedestrians, bicycles and vehicles, approaching from the right.

d)   Pedestrians, bicycles and vehicles, approaching from the left.

e)   None of the above.

11.   When you are driving faster than other traffic on a freeway, which lane should you use?

a)   The right lane.

b)   The shoulder.

c)   The left lane.

d)   The carpool lane.

e)   None of the above.

12.   When you see a car approaching on your lane you should:

a)   Pull to the right and slow down.

b)   Sound your horn.

c)   At night, flash your lights.

d)   All of the above.

e)   None of the above.

Test on Safe Driving Practices/Defensive Driving

1.   In your opinion, which of the following is the best example of defensive driving?

a)   Keeping an eye on the car's brake lights in front of you while driving.

b)   Keeping your eyes moving looking for possible hazards.

c)   Putting one car's length between you and the car ahead of you.

d)   To check your mirrors a couple of times each trip.

2.   Swerving right instead of toward oncoming traffic to prevent a crash is better.

a)   True

b)   False

Answer: A

3. The very minimum amount of a safe following distance is?

a) 2 seconds.

b) 4 seconds.

c) 6 seconds.

d) 8 seconds.

4. Which of the following acts is most effective in avoiding a collision?

a) Keeping your lights on at all times.

b) Wearing a seat belt.

c) Driving in daytime hours only.

d) To Keep a cushion of space at all times.

e) Driving slow at all times.

5. Which of the following emotions would have a significant effect on your ability to drive safely?

a) Worried

b) Excited

c) Afraid

d) Angry

e) Depressed

f) All of the above

g) None of the above

6. Signaling in advance and keeping a steady speed when you have to slow down or turn will help maintain what

a) A safe distance ahead of your vehicle.

b) A safe distance behind your vehicle.

c) A safe distance next to your vehicle.

d) All of the above.

7. You must always stop at a minimum of ____ feet from a stopped school bus with its red lights flashing?

a) 10 feet.

b) 20 feet.

c) 30 feet.

d)   50 feet.

8.   Which of these practices is not only dangerous but illegal to do while driving?

a)   Adjusting your outside mirrors manually.

b)   Wearing headphones that cover both ears.

c)   Putting on make-up.

d)   Eating and drinking coffee.

e)   Reading a map

9.   If any driver is following you too closely you should?

a)   Slowly speed up.

b)   Jam the brakes.

c)   Flash your brake lights 3 times.

d)   Move over to another lane when there is room.

e)   None of the above.

10.   What is the acceptable minimum safe following distance under most conditions?

a)   A minimum of 2 seconds is the most recommended following distance under most conditions.

b)   A minimum of 4 seconds is the most recommended following distance under most conditions.

c)   A minimum of 5 seconds is the most recommended following distance under most conditions.

d)   A minimum of 6 seconds is the most recommended following distance under most conditions.

e)   None of the above.

11.   If you are at a highway entrance and has to wait for a gap in traffic before entering the roadway, what would you do?

a)   Pull up as far as possible on the ramp and wait for some room behind you on the ramp for other vehicles.

b)   Drive to the shoulder, wait for a gap in the roadway, then accelerate quickly.

c)   I would slow down at the entrance ramp and wait for a gap, speed up to enter at the same speed that traffic is moving.

d) At the entrance ramp slow down and sound your horn and activate your emergency flash light

e) None of the above.

12. When you see a car approaching on your lane you should:

a) Pull to the right and slow down.

b) Sound your horn.

c) At night, flash your lights.

d) All of the above.

e) None of the above.

13. It is legal to park next to a fire hydrant as long as you move your vehicle if necessary.

a) True

b) False

14. In which of the following conditions do you need extra following distance?

a) When driving on slippery roads.

b)   When following a motorcycle.

c)   When following trucks or vehicles pulling trailers.

d)   When it is hard to see.

e)   All of the above.

f)   None of the above.

15.   You should always avoid placing infants or a small child in the front seat of a vehicle with airbags.

a)   True

b)   False

16.   Keeping your eyes always locked straight ahead is a good defensive driving practice.

a)   True

b)   False

17.   When exactly should safety belts be worn?

a)   At all times as a passenger and as driver.

b)   Only when driving on curvy roads.

c) Only when riding in the back seat.

d) Only when driving on the freeway.

e) None of the above.

18. In which of these situations should one use horn?

a) To tell a vehicle to get out of your way.

b) To warn bikers that you are passing.

c) When changing lanes quickly.

d) To prevent a possible accident.

e) To let someone know you are angry.

f) All of the above.

Test on Special Driving Situation

1. In which of the following situations should horn be used?

a) In which of the following situations should horn be used?

A) To notify a vehicle to move out of the way.

b) To warn bike riders that you are passing.

c) When changing lanes quickly.

d) To prevent a possible accident.

e) To let someone know you are angry.

2. At what point is the road most slippery when raining?

a) When it first starts to rain.

b) Only after it has been raining for a while.

c) After the rain has stopped, but the road is still wet.

d) None of the above.

3. Name some of the places where you are likely to find slippery spots on the road?

a) In corners and at stop signs.

b) shady spots and on overpasses and bridges.

c) In tunnels and on hills.

d) Near large bodies of water.

4. One of the basic important things to remember about driving at night or in a fog is to?

a) Be ready to brake more quickly.

b) Watch for cars at intersections.

c) Drive within the range of your headlights.

d) Use your high beams at all times.

e) None of the above.

5. In which of these conditions would you need an extra following distance

a) When driving on slippery roads.

b) When following a motorcycle.

c) When following trucks or vehicles pulling trailers.

d) When it is hard to see.

e) None of the above.

f) All of the above.

6. When driving in the snow or rain during the day you should?

a) Use your high beams.

b) Use your fog lights.

c) Use your low beams.

d) Use no headlights.

7. During a skid, one should steer to the left if the rear of the vehicle is skidding in what direction?

a) The left.

b) The right.

8. The most important thing to always remember when controlling a skid is to apply the brakes firmly

a) True

b) False

9. What should you do if you are on the highway entrance and have to wait for a gap in traffic before entering the roadway?

a) Pull up on the ramp as far as you can and wait leaving some room behind you on the ramp for other vehicles.

b)   Drive quickly to the shoulder and wait for a gap in the roadway, then accelerate quickly.

c)   Slow down on the entrance ramp and wait for a gap, then speed up so you enter at the same speed that traffic the is moving.

d)   Sound the horn and activate your emergency flashing lights to alert drivers that you are entering the roadway.

10)   How far ahead should look when you are driving in town?

a.   1 city block.

b.   1 quarter mile.

c.   1 half mile.

d.   1 mile.

11.   When you experience glare from a vehicles headlights at night you should?

a)   Look above their headlights.

b)   Look below their headlights.

c)   Be Looking towards the right edge of your lane.

d) Be Looking toward the left edge of your lane.

12. Whenever you drive in a heavy fog during the daytime you should drive with your

a) Headlights off.

b) Parking lights on.

c) Headlights on low beam.

d) Headlights on high beam.

e) None of the above.

13. Whenever you drive on slippery roads, you should increase your following distance by _____

a) 2 times.

b) 3 times.

c) 4 times.

d) 5 times.

14. You must use high beam lighting during heavy rain, in fog, and snow.

a) True

b) False

15. During a heavy rain, you can lose all traction and start hydroplaning at?

a) 25 mph

b) 40 mph

c) 50 mph

d) 65 mph

16. Safety belts would help you keep control of your car?

a) True

b) False

17. On a rainy, foggy, or snowy day when it starts to get dark, and   if driving off a rising or setting sun, it is a very good time to?

a) Check the tires.

b) Put on your seatbelt.

c) Turn on your headlights.

d) Roll up the windows.

18. If your vehicle suddenly begin to skid you should?

a) Turn the steering wheel in the direction that you desire the vehicle to go, then, use the brake,

b) Avoid the brake and turn the steering wheel in the direction that you want the vehicle to go.

c) Use the brake and turn the steering wheel in the direction that you do not want the vehicle to go.

d) Avoid the brake and turn the steering wheel in the direction you do not want the vehicle to go.

19. You may turn your vehicle while braking with ABS with less or no skidding than with regular brakes.

a) True

b) False

20. On a rainy, a snowy, or foggy day when it begins to get dark, and when you gradually drive away from a rising or setting sun, it is a good time to?

a) Check the tires.

b) Put on your seatbelt.

c)   Turn on your headlights.

d)   Roll up the windows.

e)   None of the above.

21.   Whenever driving in heavy fog during the daytime you should drive with your?

a)   Headlights off.

b)   Parking lights on.

c)   Headlights on low beam.

d)   Headlights on high beam.

e)   None of the above.

Test on Vehicle Positioning

1.   It is legal to park next to a fire hydrant as long as you will move your vehicle if necessary

a)   True

b)   False

2.   When you park on an uphill grade, which way should you turn your wheels

a) Left

b) Right

c) Straight

3. How close should one pack when parking next to a curb?

a) Not closer than 6 inches from the curb.

b) Not farther than 6 inches from the curb.

c) Not closer than 12 inches from the curb.

d) Not farther than 18 inches from the curb.

e) Not farther than 24 inches from the curb.

4. Whenever you park uphill on a street with no curb, which way should your front wheels be turned?

a) To the left.

b) To the right.

c) Parallel with the road.

d) None of the above

5. In which of these places must you never park

a) On sidewalks.

b) In bicycle lanes.

c) In front of driveways.

d) By fire hydrants.

e) All of the above.

Test on Drugs and Alcohol Consumption

1. What drug can affect your ability to drive safely?

a) Almost every drug prescription or over the counter drugs can affect your ability to drive.

b) Alcohol and marijuana.

c) Only illegal drugs.

d) None of the above.

2. How many drinks could it take to affect your driving

a) 1

b) 2

c) 3

d) 4

e) 5

3. After drinking, a cold shower or coffee will lower your blood alcohol content.

a) True

b) False

4. Which of the following influences the effects of alcohol in the body?

a) The time between each drink.

b)   The body weight of a person.

c)   The amount of food in the stomach.

d)   All of the above.

e)   None of the above.

5.   Of a truth, your judgment and vision are both affected after drinking alcohol. Which is affected first?

a)   Judgment

b)   Vision

Test on Teen Driver Safety

1.   Which of the following is the leading cause of death for teens in the US?

a)   Cancer

b)   Suicide

c) Auto crashes

d) Murder

2. 16 to 19 year teenagers are __ times more likely than drivers 20 and over to be in a fatal crash

a) 0

b) 2

c) 3

d) 4

3. 16 to 17 year old teenage driver fatality rates decrease with each additional passenger added to a vehicle.

a) True

b) False

4. In 2008 63 percent of teenage passenger deaths occurred in vehicles driven by another teenager.

a) True

b) False

5. In 2011, how many high school teens drank alcohol and drove?

a) 1,000 teens.

b) 10,000 teens.

c) 100,000 teens.

d) 1,000,000 teens

6. Which days of the week does 55 percent of teen driving deaths occur?

a) Monday and Tuesday

b) Wednesdays

c) Thursdays

d) Friday, Saturday, and Sunday

7. The vehicle death rate for male teenage drivers and passengers is almost twice that of females.

a) True

b) False

8. The age group that is 3 times more likely to die in a motor vehicle crash than the average of all other drivers combined, is?

a) 16

b) 17

c) 18

d) 19

9. The risk of vehicle crash is higher for which age group over all other groups?

a) 16 to 19

b) 20 to 22

c) 23 to 25

d) Age does not matter

10. In 2010, 22 percent of drivers between the ages of 15 and 20 years involved in fatal crashes were drinking

a) True

b) False

11. Teens have the lowest rate of using seat belt compared with other age groups.

a) True

b) False

12. Will you share this quiz with a teen you care about?

a) Yes

b) No

13. Two hundred and eighty two thousand teens were injured in vehicle crashes in 2010

a) True

b) False

14. In 2008 what percent of teenage motor vehicle crash deaths were passengers in the vehicles?

a) 22

b) 37

c) 56

d) 81

Test on Driving an Intersection &Turns

1. A driver turning left at an intersection should yield to which of the following?

a) Vehicles approaching from the opposite direction.

b) Pedestrians, bicycles and vehicles approaching from the opposite direction.

c) Pedestrians, bicycles and vehicles approaching from the right.

d) Pedestrians, bicycles, and vehicles approaching from the left.

e) None of the above.

2. The same general rules apply as for maneuvering through any other type of intersection when driving at a roundabout

a) True

b)   False

3.   If more than one vehicle is stopped at an intersection, which vehicle has the right-of-way?

a)   The largest vehicle.

b)   The first vehicle that attempts to go.

c)   The first vehicle to arrive.

d)   The vehicle on the right.

e)   None of the above.

4.   Whenever you want to turn left at an intersection but oncoming traffic is heavy. What should you do?

a)   Wait at the crosswalk for traffic to clear.

b)   Wait right in the center of the intersection for traffic to clear.

c)   Start your turn forcing others to stop.

d)   None of the above.

5.   While waiting at the intersection to complete a left turn, you should

a)   Sound your horn so vehicles will let you get through.

b)   Keep your wheels turned to the left while signaling and waiting for an opening.

c)   Keep your wheels turned straight, signal and wait for an opening.

d)   Put on your headlights.

6.   When turning left from a two-way street to a one way street, your vehicle should be in which lane when the turn is completed

a)   In the right lane.

b)   In the left lane.

c) In the center lane.

d) None of the above.

7. How many feet should you signal before your intended turn?

a) 25 feet.

b) 50 feet.

c) 75 feet.

d) 100 feet.

e) 200 feet.

8. It's ok to pass when approaching a hill-top or a curve if you hurry?

a) True

b) False

9. If you have entered an intersection already when the light changes, you should?

a) Stop in the intersection.

b) Proceed and clear the intersection.

c) Flash your lights through the intersection.

d) Sound your horn through the intersection.

10. What would you do if you are at an intersection and you hear a siren?

a) Stop and do not move until every emergency vehicle has passed.

b) Continue through the intersection, and pull over to the right side of the road and stop.

c) Back up then, pull over to the right side of the road and stop.

d) None of the above

11. You have to yield to traffic on your left already in the roundabout.

a) True

b) False

12. Unless prohibited by a sign, when can one turn left at a red light?

a) In an emergency.

b) Turning from a two way road to a one way road.

c) Turning from a one-way road to a one way road.

d) Never.

13. A driver who approaches an intersection should yield the right-of-way to traffic that is at the intersection.

a) True

b) False

Test on Visuals

1. Scanning and seeing events very well in advance will help prevent what?

a) Fatigue

b) Distractions

c) Panic stops

d) Lane changes

2. Night driving is said to be more dangerous because?

a) Traffic signs are less visible at night.

b) The distance one can see ahead is reduced.

c) People are sleepy at night.

d) Criminals come out at nighttime.

e) None of the above.

3. How can you see when there is a car in your blind spot?

a) Lean backward and forth looking in your mirrors.

b) Look over your shoulder.

c) Adjust the power mirrors if you have them.

d) Nothing can be done, that is why it is called a blind spot.

4. When you are on the highway, how far ahead should you look?

a) 3 to 5 seconds.

b) 5 to 10 seconds.

c) 10 to 15 seconds.

d) 15 to 20 seconds.

5. What should you do if you are on the highway entrance and have to wait for a gap in traffic before entering the roadway?

a) Pull up on the ramp as far as you can and wait leaving some room behind you on the ramp for other vehicles.

b) Drive quickly to the shoulder and wait for a gap in the roadway, then accelerate quickly.

c) Slow down on the entrance ramp and wait for a gap, then speed up to enter at the same speed of traffic

d) Slow down at the entrance ramp and wait for a gap, then sound the horn and put-on the emergency flash lights

6 How far ahead should you look when you are driving in town?

a. 1 city block.

b. 1 quarter mile.

c.   1 half mile.

d.   1 mile.

7.   When you experience glare from a vehicles headlights at night you should?

a)   Look above their headlights.

b)   Look below their headlights.

c)   Be Looking towards the right edge of your lane.

d)   Be Looking toward the left edge of your lane.

8.   Being awake 24 hours may cause impairment that is nearly equal to that of an alcohol content of what?

a)   .02

b)   .05

c)   .08

d)   .10

Answer: D

9.   Besides helping you see at night, headlights help other road users see you at any time.

a)   True

b)   False

Answer Section

Test on General Driving Experience

1. Answer: D   21Answer: D   41 Answer: B   61Answer: D

2. Answer: C   22 Answer: C   42 Answer: B   62 Answer: A

3 Answer: A   23Answer: B   43 Answer: A   63Answer: B

4 Answer: D   24 Answer: A   44Answer: B   64Answer: B

5 Answer: B   25 Answer: D   45Answer: A   65Answer: D

6 Answer: B 26 Answer: A 46Answer: C 66 Answer: A

7 Answer: C 27 Answer: D 47Answer: C

8 Answer: D 28 Answer: C 48 Answer: D

9 Answer: D 29 Answer: C 49 Answer: B

10 Answer: A 30 Answer: A 50 Answer: B

11 Answer: B 31 Answer: A 51 Answer: B

12 Answer: B 32 Answer: A 52Answer: A

13 Answer: D 33 Answer: A 53 Answer: B

14 Answer: A 34 Answer: B 54 Answer: E

15 Answer: D 35 Answer: B 55 Answer: C

16 Answer: C 36 Answer: B 56 Answer: B

17 Answer: B 37 Answer: A 57 Answer: D

18 Answer: D 38 Answer: A 58 Answer: D

19 Answer: B 39 Answer: A 59 Answer: B

20 Answer: C 40 Answer: B 60 Answer: B

Test on Knowledge of Road Signs &Traffic Control

1. Answer: A 21 Answer: C 41 Answer: E 61Answer: D

2. Answer A 22 Answer: B 4 Answer: C   62 Answer: B

3 Answer: C   23 Answer: B 43 Answer: A 63 Answer: D

4 Answer: C   24 Answer: D 44 Answer: C 64Answer: D

5 Answer: A   25 Answer: D 45 Answer: A 65 Answer: C

6 Answer: B   26 Answer: B 46 Answer: C

7 Answer: C   27 Answer: E 47Answer: C

8 Answer: C   28 Answer: F 48 Answer: C

9 Answer: C   29 Answer: B 49 Answer: D

10 Answer: B 30 Answer: C 50 Answer: D

11 Answer: B 31 Answer: C 51 Answer: D

12 Answer: D 32 Answer: C 52 Answer: B

13 Answer: B 33 Answer: B 53 Answer: B

14 Answer:C 34 Answer: C 54 Answer: A

15 Answer: A 35 Answer: A 55 Answer: B

16 Answer: B 36 Answer: C 56 Answer: C

17 Answer: C 37Answer: E 57Answer: C

18 Answer: B 38 Answer: E 58 Answer: B

19 Answer: B 39 Answer: B 59 Answer: D

20 Answer: A 40Answer: A 60Answer: E

Test on Lane Control

1.Answer: B   2 Answer: B   3Answer: A   4 Answer: C

5Answer C   6 Answer: D   7 Answer: B   8 Answer: B

9Answer: A   10 Answer: B 11Answer: C   12 Answer: D

Test on Safe Driving Practices/Defensive Driving

1.Answer: B   7Answer: B   13Answer: B

2. Answer A   8Answer: B   14 Answer: E

3 Answer: B   9 Answer: D   15 Answer: A

4 Answer: D   10 Answer: B 16 Answer: B

5 Answer: F   11 Answer: C 17 Answer: A

6Answer: B   12Answer: D   18Answer: D

Test on Special Driving Situation

1. Answer: D   7 Answer: A   13 Answer: A 19 Answer: A

2. Answer A   8 Answer: B   14 Answer: A 20 Answer: C

3 Answer: B 9 Answer: C   15 Answer: C 21 Answer: C

4 Answer: C   10 Answer: A 16 Answer: A

5 Answer: F   11 Answer: C 17 Answer: C

6 Answer: D   12 Answer: C 18 Answer: B :

Test on Vehicle Positioning

1. Answer: B   2 Answer: A   3 Answer: D   4 Answer: B

5. Answer E

Test on Drugs and Alcohol Consumption

1. Answer: A   2 Answer: A   3 Answer: B   4 Answer: D

5. Answer A

Test on Teen Driver Safety

1.Answer: C	5Answer: D	9Answer: A	13Answer: A
2.Answer C	6Answer: D	10Answer: A	14Answer: D
3Answer: B	7Answer: A	11Answer: A	
4Answer: A	8Answer: A	12Answer: A	

Test on Driving an Intersection &Turns

1. Answer: A 2 Answer: A 3 Answer: C 4Answer: B

5. Answer C 6 Answer: B 7 Answer: D 8 Answer: B

9 Answer: B 10 Answer: B 11Answer: A 12 Answer: C

13Answer:A

Test on Visuals

1. Answer: C   4Answer: C   7Answer: C

2. Answer B   5 Answer: C   8 Answer: D

3 Answer: B   6 Answer: A   9 Answer: A

www.ingramcontent.com/pod-product-compliance
Lightning Source LLC
Chambersburg PA
CBHW021447210526
45463CB00002B/664